Golden ratio in the elliptical honeycomb

Daniel Favre - *e*phiscience

Ne faire fi de rien
Pour faire Phi de tout

Préface en langue française de l'ouvrage de Daniel Favre.

Par Léonard Ribordy.

Le **Nombre d'or** est connu sous sa forme de proportion géométrique depuis les premières dynasties égyptiennes. Il a servi à élaborer les proportions exceptionnelles de la Grande pyramide dite de Khéops et tous les temples dédiés aux multiples divinités qui peuplent la longue vallée du Nil.

Cette connaissance passa au monde grec puis romain pour imprégner durablement l'occident chrétien.

Ce n'est qu'au XIIe siècle de notre ère chrétienne que, grâce aux travaux de Léonardo di Pisa dit Fibonacci ce Nombre exceptionnel trouva sa définition arithmétique, sous la forme d'une série de chiffres.

1 1 2 3 5 8 13 21 34 55 89 144 233 377 etc.

Il est à relever que cette série conduit au Nombre d'or en affinant sa perfection qui ne se révèle qu'à l'infini de la série. Cette remarque sous-entend que les premiers chiffres de la série ne dévoilent pas le Nombre d'or dans la rigueur de sa précision arithmétique. Ce n'est qu'à partir du 12ème étage de la série que l'on voit le Nombre d'or se révéler avec une précision relative, mais suffisante ; 144/89 = 1,61797 ≈ 1,61803.

Cela refroidit les certitudes des idéalistes qui considèrent que le Nombre d'or est omniprésent dans le développement des plantes, de certains légumes et autres graines.

En effet toutes les mesures le confirment, la nature se développe avec des chiffres se situant en dessous du 12ème étage de la série.

Donc la Nature recherche le Nombre d'or mais sans jamais l'atteindre dans sa perfection.

Il en va de même pour les proportions du corps humain, de la tête et d'autres parties du corps.

Seuls des artistes comme Léonard de Vinci, Albrecht Dürer et plus près de nous, Matila Ghyka ont pu dessiner des corps construits selon les proportions idéales, qui n'existent dans aucun corps humain des deux sexes, fut-il canon de beauté genre poupée Barbie.

Ces remarques n'enlèvent rien à la valeur exceptionnelle du célèbre Nombre d'or. Après avoir été identifié comme proportion géométrique idéale pour créer l'harmonie, avec Fibonacci (1175-1240) ce Nombre acquiert la valeur de l'or: **Nombre d'or,** car ses propriétés géométriques et arithmétiques, sont absolument époustouflantes et ont été identifiées au cours des âges sans être pour autant exhaustives, car on n'est pas encore au bout des découvertes. Deux siècles après Fibonacci un autre mathématicien italien, ami de Léonard de Vinci nommé Luca Pacioli (1445-1510) mit au point une autre série de chiffres conduisant au Nombre d'or, mais avec la même restriction que celle de son illustre prédécesseur Fibonacci; de nouveau, les premiers étages de sa série ne dévoilent pas le Nombre d'or dans toute sa perfection.

Par contre Luca Pacioli attribua à ce Nombre un caractère divin: **la Divine proportion**; il consigna ses découvertes dans un ouvrage portant ce titre.

Toutes les analyses que l'on peut faire avec les multiples propriétés de ce Nombre confirment sa valeur de **Divine proportion**. La philosophie du sacré attribue à ce Nombre la valeur de la **Vie divine.** Celle qui gouverne dans l'ensemble de l'Univers, le phénomène engendrant le vivant sur certaines planètes telluriques bien disposées à recevoir ce trésor: **La Vie.**

Mon objectif littéraire n'est pas de convaincre ni de convertir personne: cependant mes convictions personnelles ont été à tel point renforcées par mes découvertes au cours de trente-cinq années de recherches liées au Nombre d'or que je les ai consignées dans trois livres qui furent publiés aux Editions Trajectoire à Toulouse, entre 2007 et 2014.

Pour en venir à la publication de Monsieur Daniel Favre qui pour son article consacré au développement des rayons des abeilles m'a demandé cette préface, je retrouve ce que j'ai moi-même observé en tant que fils et petit-fils d'apiculteur.

Les abeilles ont été des modèles de précisions géométriques dans l'art de construire les alvéoles hexagonales parfaitement adaptées à leurs besoins de réaliser des rayons de cire porteurs de larves, de pollen et de miel. La **nature** les a induites à savoir disposer ces alvéoles pour être auto-stables, inclinées juste ce qu'il faut pour permettre à leur contenu de ne pas s'échapper sans contrôle.

Comme ingénieur, je suis en admiration par la façon dont les rayons porteurs des alvéoles sont ancrés et portés par les branches ou des listes de bois auxquelles le poids du rayon chargé de miel et d'abeilles n'affecte pas la résistance. Comment par le miracle chimique de la propolis ces rayons sont collés et auto-protecteurs contre les divers prédateurs.

Il est vrai que la forme naturelle de ces rayons peut épouser la forme d'une ellipse construite à l'intérieur de la surface fermée d'un rectangle à double carré (1 sur 2), dit carré-long ou d'un autre épousant la forme d'un rectangle d'or (1 sur Φ) ce qui laisse supposer que les abeilles possèdent instinctivement les vertus métaphysiques du Nombre d'or puisque ces deux formes en sont le plus beau reflet.

Une remarque de l'auteur lui-même relativise cette observation quand il dit ...*in the biological world, living systems are always only approaching their exact mathematical model*. Ce qui signifie dans notre langue que ...dans le monde biologique, le système vivant ne fait qu'approcher l'exact modèle mathématique.

Cette remarque tempère donc l'idéal proposé par les abeilles, qui comme tout ce qui constitue le règne du vivant, recherche le Nombre d'or mais sans jamais l'atteindre dans sa perfection mathématique. Laissons donc aux abeilles la liberté de développer comme bon leur semble leur habitat, même en dehors de la perfection de l'ellipse, en suivant les lois de l'évolution.

Comme tout ce qui vit développe l'**intelligence**, (qui est la faculté de s'adapter à son environnement), peut-être que les abeilles comme nous, savent-elles intuitivement que l'intelligence conduit à la prise de **conscience**, qui elle-même conduit à la **connaissance.**

Léonard Ribordy, Ingénieur civil EPFL
Le 21 février 2016

La Divine proportion par la géométrie et les Nombres: Ed. Trajectoire 2007
Architecture et géométrie sacrées dans le monde, à la Lumière du Nombre d'or: Ed Trajectoire 2010
Vers l'équation de Dieu, par Oranda l'âme de la Vie: Ed Trajectoire 2014

Vorwort von Hochwürden Paul Probst

Der Anfrage von Dr. Daniel Favre zu dieser wissenschaftlichen Arbeit ein Vorwort zu schreiben, bin ich nicht nur sehr gerne nachgekommen, sondern empfinde es als eine Ehre. Wenn Dr. Favre im Sinne einer Schlussfolgerung konstatiert, dass die mathematische Präsenz des Goldenen Schnittes möglicherweise "die Wirkung eines inhärenten Gesetzes des Kosmos offenbart", so gibt er damit der Naturwissenschaft einen neuen Ausblick, der sogar über das hinausgeht, was als Paradigmenwechsel bezeichnet wird.

Daniel Favre untersucht in dieser Arbeit akribisch die frühen Stadien des Aufbaus der Bienenwaben und stellt fest, dass deren elliptische Form nicht zufällig ist, sondern mathematischen Regeln folgt, welche an einer Geometrie orientiert sind, die in einem engen Bezug zum Goldenen Schnitt steht.

Folgende Frage stellt sich in diesem Zusammenhang: Wie ist es den Bienen möglich, ein solches Muster zu erkennen, und ihre Waben nach Massgabe dieses Musters beziehungsweise dieser Ordnung zu bauen? Üblicherweise wird hier auf den Instinkt der Bienen verwiesen. Damit ist allerdings noch nichts erklärt. Woher kommt der "Instinkt"? Wer setzt den Impuls? Woher kommt das Muster, dem gemäss die Waben gebaut werden? Wer oder was hat dieses Muster kreiert und im Sinne einer offensichtlich speziesverbindlichen Ordnung festgelegt ?

Der Schluss liegt nahe, dass die Bienen auf welche Art auch immer in ihrem Tun gelenkt werden, was ihnen erlaubt, einem übergeordneten Plan zu folgen. Die Natur ist hierarchisch aufgebaut und jeder Teil davon kann sich im Rahmen der Evolution weiterentwickeln, wobei die Evolutionen der Schöpfung immer auch in einer allumfassenden Göttlichen Ordnung aufeinander bezogen sind. Diese Ordnung ist nicht als starres Gebilde, sondern wie alles Leben als in Entwicklung begriffen zu verstehen. Das Göttliche, Gott, wird hier allerdings nicht als ein weiser, alter Mann mit grauem Bart verstanden. Solch ein Gott existiert nicht.

Der Universale GOTT, von dem hier die Rede ist, ist eins, doch "Er" ist mehr als eins; ALLE GUTEN DINGE SIND GOTT; alle Dinge sind eins, aufeinander bezogen und miteinander verbunden. In jeder Zelle, jedem Molekül, Atom und Elektron gibt es Leben. Sogar im winzigsten Partikel Erde gibt es das Leben GOTTES. Kurz: Es gibt auf Erden nichts dergleichen wie ein unbelebtes Objekt; dies bestätigt auch die Quantenphysik. Und alles Leben ist Teil einer grösseren, umfassenden Ordnung.

Wenn Daniel Favre folgert, dass die mathematische Präsenz des Goldenen Schnittes möglicherweise "die Wirkung eines inhärenten Gesetzes des Kosmos offenbart", so kann ich dem aus einer geisteswissen-schaftlichen Perspektive vollumfänglich beipflichten. Eine Göttliche Ordnung durchdringt die ganze Schöpfung. Ohne eine solche Ordnung wäre Leben, wie wir es kennen, unmöglich. Ob im Wunderwerk des menschlichen Körpers oder in den unendlichen

Galaxien des Universums, alles ist in eine Ordnung eingebunden. Wenn Elemente nicht mehr mit dieser Ordnung verbunden sind, wird der gesamte Organismus krank. Im Falle des menschlichen Körpers reden wir dann von Krebs Zellen, die nicht mehr in diese Ordnung eingebunden sind und deshalb dysfunktional wachsen, wuchern und den ganzen Körper im wahrsten Sinne des Wortes in "MitLeidenschaft" ziehen.

Es ist in der Tat so, dass die gesamte Natur, von Kleinstlebewesen wie Mikroben bis hin zum Menschen der Krone der Schöpfung in eine Ordnung eingebunden ist. Was uns Menschen von den Tieren unterscheidet, ist die Freiheit der Wahl. Wir können uns (zeitweilig) gegen diese Ordnung wenden. Tatsächlich haben wir uns sehr weit vom Göttlichen und damit der Göttlichen Ordnung entfernt. Der katastrophale Zustand des Planeten Erde, unseres Heims, zeugt davon.

Betrachtet man diese Ordnung und ihre Genialität, wie alles Leben, insbesondere auch die vier Elemente, aufeinander bezogen ist, so dass dieser Planet und das Leben auf ihm möglich ist, so könnte man in der Tat von einem "Wunder" sprechen. Aber: Macht der Begriff "Wunder" Sinn? Ja und nein. "Wunder" in dem Sinn, dass die von Dr. Favre erforschten Zusammenhänge auf eine der Natur inhärente Göttliche Ordnung verweisen, über die wir uns in Demut wahrlich "wundern" können, ja. Allerdings verwenden wir den Begriff "Wunder" oft, um damit Dinge oder Phänomene zu benennen, die wir uns aus unserer sehr begrenzten Sicht einfach nicht erklären können.

Mit einem Grund für unser sehr beschränktes Verständnis von Zusammenhängen ist das überwiegend mechanistische Weltbild, das uns nicht nur von einer philosophischen Betrachtungsweise der Welt entfernt, sondern auch unserer geistigen Beziehung zur Natur ein Ende gesetzt hat. Wir leben in der Tat in einer weitgehend geistlosen Zeit. Die rein empirisch orientierte Naturwissenschaft erklärt uns die Welt und ihre Zusammenhänge auf der Grundlage eines materialistischen Weltbildes. Gott kommt darin nicht vor, weil es dafür keinen empirischen Beweis gibt. Allerdings bleibt es nicht bei dieser Aussage, sondern viele Wissenschaftler gehen noch weiter und folgern: Da es keinen empirischen Beweis für die Existenz Gottes gibt, existiert Gott auch nicht. Dabei wird übersehen, dass die Unmöglichkeit, ein Phänomen empirisch oder messbar zu erfassen, nicht bedeutet, dass es nicht existiert. Auch wenn es beispielsweise während Jahrhunderten keine wissenschaftlichen Instrumente gab, welche Radioaktivität messen konnten, existierte sie trotzdem auch schon zu diesem Zeitpunkt. GOTT ist kein Ding, ebenso wenig wie ein Gedanke eines ist. Meines Wissens hat zum Beispiel noch kein Gehirnscan je vermocht, einen Gedanken zu erfassen. Auch "Geist" oder "Liebe" sind nicht messbar, aber kaum jemand würde wohl behaupten, dass sie nicht existieren.

In der heutigen Zeit wissen viele Menschen nicht, woher sie kommen und wohin sie gehen, und sie haben den Kontakt zu sich selbst verloren. Um dies zu korrigieren und die richtigen Antworten zu finden, muss der Mensch wieder lernen, in sich hineinzuhorchen und sich mit seiner eigenen Göttlichkeit zu verbinden. Die Spaltung zwischen Materie und Geist muss auch in der Wissenschaft aufgehoben werden. Die Naturwissenschaft muss sich wieder mit der Geisteswissenschaft vereinen. Einstein drückte es so aus: "Wissenschaft ohne Religion ist lahm, Religion ohne Wissenschaft ist blind." Das Leben ist nicht auf den physischen Ausdruck beschränkt. Es zu verstehen, verlangt einen erweiterten Horizont. Was wir physisch sehen, ist nur ein Teil der Realität. Neben der Materie gibt es auch einen feinstofflichen Bereich, den wir, wenn wir nicht selbst über die entsprechenden Fähigkeiten verfügen, zum Beispiel mittels der Kirlianfotografie sichtbar machen können. Auch die Intuition, die Einstein als "göttliches Geschenk" bezeichnete, ist nicht sichtbar. Verschiedene geistige Richtungen nennen sie die "Stimme der Seele", das Bindeglied zwischen Körper und Geist und die Verbindung zwischen dem Individuellen und dem Universellen. Die Eingebungen, die aus dieser Verbindung resultieren, sind oft nicht sprachlicher Natur.

So hat Einstein die Relativitätstheorie als inneres Bild empfangen und anschliessend versucht, sie anhand der (menschlichen) Mathematik zu erfassen. Mozart hat seine Symphonien in ihrer Gänze als Bild gesehen und sie erst nachher in Form von Noten zu Papier gebracht.

Viele Genies haben durch eine solche bildhafte Verbindung zu ihrer Göttlichkeit im Inneren ihre bahnbrechenden Einsichten empfangen.

Aber dies ist nicht nur Genies vorbehalten. Viele Kleinkinder sind dazu fähig, verdrängen dies aber, wenn sie älter werden, weil es viel Mut braucht, um das, was man sieht, in der heutigen Welt auch auszudrücken, mitzuteilen und sich damit zu exponieren. Aber wir alle können es wieder lernen. Die Verbindung zum Göttlichen ist unser Geburtsrecht es liegt an uns, dieses wieder zu beanspruchen.

The Very Reverend Dean Paul Probst,
SF em Europäischer Präsident
The World Foundation for Natural Science
22. Februar 2016

Golden ratio (*Sectio aurea* ; divine proportion) in the early elliptical honeycomb of *Apis mellifera*

Daniel Favre
Dr. phil. nat.

Ephiscience   www.ephiscience.net

Biologist, independent researcher, retired from academic institutions.

Email: info@ephiscience.net

Short title :
Elliptical honeycomb and golden mean

Favre D (2016) Golden ratio (*Sectio Aurea*) in the Elliptical Honeycomb.
Journal of Nature and Science (JNSCI), 2(1):e173.
http://www.jnsci.org/content/173

Abstract

The honeybee comb, which is highly similar among honeybee species, is a mass of six–sided cells made by honeybees. It contains the brood, the honey and the pollen within horizontally–arranged and parallel structures. The construction processes and the geometry of the hexagonal cells have been extensively studied since centuries. Although studies of the natural, full–sized comb structure have been thoroughly performed in the past, the analysis of their early developmental stage size properties has not been investigated in detail. Here in particular, I found that the general two–dimensional elliptical form of the newly–constructed honeycombs could be drawn into a rectangle of modules having values approaching either 2.00 or 1.62, where the module of the rectangle is the simple division of it's long by it's short side lengths. It is proposed here that the elliptical form of the early stage honeybee comb is not random, but is following mathematical rules reflecting some geometry intimately related to the golden ratio, also called golden mean or divine proportion. It is proposed that the elliptical honeycomb is showing an intrinsic gnomonic growth. This mathematical presence of the golden ratio might reveal the effect of an inherent law of the Cosmos.

Introduction

A honeycomb is certainly one of the most beautiful and impressive natural structures. It's study is not only allowing the understanding of the honeybees' natural conditions of living, but it is also a fascinating piece of natural wax architecture. The honeycomb is a crucial part of the honeybee's nest, whose composition, structure and function have been extensively reviewed [1]. An exhaustive coverage of the synthesis and secretion of beeswax, its elaboration into combs and the factors that bear on the execution of these processes by honeybees has been reviewed elsewhere [2]. In particular, the construction of the hexagonal cells and the regulation of the space between adjacent combs have been a matter of extensive research [3]. Analyses of freely–built combs in mixed or pure *A. mellifera* and *A. cerana* (Hymenoptera, Apoidae, Apidae) colonies, in terms of numbers of festoons, number of honeybee workers on festoons, percentage of irregular cells, cell size and patterns of newly built combs, have been presented elsewhere [4]. The biological foundations of swarm intelligence of bees, ants and locusts have been extensively reviewed; in the case of the honeybee, the pattern of the hexagonal cells in the combs can be thought to be the result of the darwinian natural selection, or the application of simple physical or mathematical rules [5]. The formation of the hexagonal pattern itself can be explained by wax flowing in liquid equilibrium, in which "the structure of the combs of honeybees results from wax as a thermoplastic building medium, which softens and hardens as a result of increasing and decreasing temperatures" [6]. These original results were supported later by independent researchers [7].

Physicians have proposed a mathematical model for honeycomb construction in which, via a set of dynamical coupled partial differential equations, the essential dynamical features of bee–bee and bee–wax interactions are integrated [8]. This domain has been the subject of intensive research (reviewed in [9]). Two main mechanisms can hypothetically explain the hidden geometry of the honeycomb, namely the diffusion-limited aggregation [10], and the constructal theory [11].

I have certainly not the perception that mathematics is a dull subject with no connection to real life. As a scientist, I am also not " studying nature just because it is useful, but because it is beautiful" [12]. Among many other scientists, I believe mathematics is nature's language. Sometimes this language is (very) complicated. Sometimes it is more simple, but not directly evident, or even hidden.

It is obvious that honeycombs do present, at peculiar stages during their growth progression, the signature or pattern of ellipses (Figure 1). This is a well–known and widely accepted fact, since the normal sketch of the comb does possess an ellipsoidal form that is typically described by all of the authors involved in the study of the beeswax construction [13,14].

Fig 1. Some characteristics of the ellipsoid and the ellipses. Ellipsoid is the three–dimensional form characteristic of the early comb found in beehives. It contains three axes, denoted x (major axis), y (middle axis) and z (minor axis). Ellipses are two–dimensional and have two perpendicular axes about which the ellipse is symmetric. These axes intersect in the middle of the ellipse. The major axis is the longest distance between antipodal points of the ellipse. The smallest distance across the ellipse is the minor axis. Conventionally, the major radius of the ellipse is denoted a, whereas the minor radius is denoted b. The minor radius is kept constant in this figure. The ratio a/b can reflect the extension of the ellipse : ratios without units and having values of 1.50 (A), 1.70 (B), 2.00 (C) and 3.00 (D) are showing increasing vertical extensions.

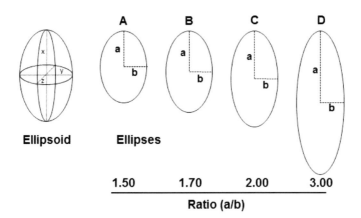

In this article, I am presenting circumstantial evidence that the early (*word emphasized*) elliptical honeycomb is based on the golden ratio. The golden ratio is an irrational number. It is represented by the Greek letter Φ or ϕ (Phi) and has the value 1.6180339887, approximately [15].

The value of Φ is calculated as (1 + √5) divided by 2. Throughout history, the golden ratio has been studied not solely by mathematicians and philosophers, but also by biologists, naturalists, artists, architects and musicians, since it was also for them an essential element for the creation and keeping of order, form and beauty [16,17]. The fascinating presence of the golden ratio in the early honeycomb is an additional stone in the edification of the Cosmos, in which it's ubiquitous presence can only be deciphered but not formally explained.

Methods

Eight healthy colonies (in Dadant–Blatt hives) were employed for honeycomb analyses that were performed between March 2015 and July 2015. In each hive, two to three frames out of a total of eleven consisted of so-called "foundationless" frames [18] allowing close–to–natural construction of the combs by the honeybees. As an apiary adviser in the Canton de Vaud (Switzerland), I followed the official beekeeping procedures provided by the Bee Research Center at the Agroscope Liebefeld–Posieux (Berne, Switzerland) [19].

During the previous autumns and winters, the *Apis mellifera carnica* honeybees have been treated against the varroa mite *Varroa destructor* with acetic acid and oxalic acid, essentially as recommended elsewhere [20]. Bee hives were located 5 km north of the city of Morges (altitude 510 m above sea level; Switzerland). In the geographic area where the study took place, the bees usually begin foraging to collect nectar and pollen in early March, depending on the weather conditions. The hives were open no more than once every seven to ten days, in order to keep the honeybees quiet most of the time. The presence of a laying queen and brood was regularly controlled.

Two main procedures were employed for the analysis of the early honeycombs. First, natural and intact honeycombs presenting evident elliptical forms were sampled for measurements using a mechanical precision caliper. The same combs were photographed using a digital camera (Panasonic DMC–TZ20 ; 3648 by 2736 pixels) followed by the analysis of the ellipses using the POWERPOINT software (Microsoft Corporation, Redmont, USA) with the usual aspect ratio of 4:3. For this, computer–generated ellipses were superimposed to each elliptical honeycomb in the picture, and the peculiar sizes (major and minor axes ; see [21]) were obtained using 200% magnification through the format calculating routine of the computer program. These numbers are given with two decimals by the computer program. Second, pictures of honeycombs were searched in both the scientific or beekeeping literature and on the internet by using a mixture of several specific keywords (such as : *Apis mellifera, dorsata, cerana,* honeycombs, colony, wax, frames, comb building & construction, foundationless frames). The digital images were used as described above, or the pictures from books were scanned using a regular printer scanner. The sizes of the elliptical honeycombs were determined electronically as described above. In order to minimize distortions in the measurements, elliptical honeycombs showing both their major and minor axes in their maximal sizes were considered. Appropriate sampling procedures were employed as objectively as possible, except for the pictures to be chosen from the occasions when the honeycombs were presenting their major outlines to the observer, i.e., when they were not observed obliquely.

The statistics reported here are the mean ± 1 standard deviation. To test for the equality of two means, the unilateral Student t-test with a heteroscedastic argument type of 3 was employed with a level of significance α = 0.01. In order to assess, whether a single extreme value could be removed from each set of data, the Dixon test was employed, with a 5% unidirectional risk. After the removal of one extreme value in each set, the normality test from Shapiro-Wilk was employed, with a level of significance of 0.01.

Results

Natural honeycombs that were built in hives containing 2 to 3 foundationless frames were found to possess evidence for a two–dimensional elliptical structure in their early stage of construction. Initial measurements of these honeycombs were performed using a mechanical caliper. This procedure is providing accurate values for the length of the middle axis of the ellipsoid, but not for its major axis, since the natural honeycomb does present a striction line and a gorge on its upper side, which is not allowing sufficiently accurate measurements. Therefore, this procedure of measurements was abandoned.

In order to obtain more accurate values with these measures, ellipses allowing the measurements of both the major and minor axes of the elliptical honeycomb were electronically drawn on the respective two–dimensional pictures. This procedure was performed with honeycombs originating from my own apiary (Fig. 2), from the beekeeping and scientific literature (Fig. 3) and from websites in the internet (Fig. 4).

Fig 2. Honeycombs found in hives containing *Apis mellifera carnica* and foundationless frames. (A) Two honeycombs found under the inner cover of the hive. The striction line and the gorge are indicated by an arrow. (B) Honeycomb built in a 10 cm x 12 cm long frame employed for comb honey production. (C), (D), (E) Honeycomb built in a foundationless frame, which has an oblique wooden bar making a plane angle of $\alpha = 26.6°$ with the horizontal bars. All honeycombs were in the brood box, except in (B) in the honey super. The ratio a/b (length of the major axis divided by the length of the minor axis) is indicated for each ellipse by a number.

Fig 3. Honeycombs found in the scientific and beekeeping literature. (A) *Apis mellifera capensis* honeycomb ([22] (© EDP Sciences, Les Ulis, France). (B), (D), (F) *Apis mellifera ligustica* honeycombs [13] (© Bernadette Darchen). (C) Early combs from honeybees [23] (Bibliothèque Nationale de France : public domain). (E) : [24] (© Springer–Verlag Berlin Heidelberg). The ratio a/b (length of the major axis divided by the length of the minor axis) is indicated for each ellipse by a number.

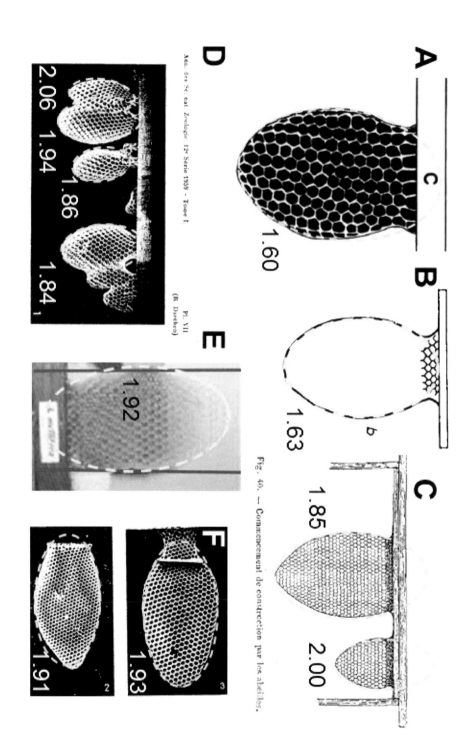

Mén. Ser. Sc. nat. Zoologie 12ᵉ Série 1959 - Tome I.

Pl. VII

(R. Darchen)

A 1.60

B 1.63 _b_

C 1.85 2.00

D 2.06 1.94 1.86 1.84₁

E 1.92

F 1.91 2 1.93 3

Fig. 40. — Commencement de construction par les abeilles.

Fig 4. Honeycomb pictures found in internet.
(A) Ottawa Honey House [25]. (B) Natural Beekeeping Trust [26] (© The Natural Beekeeping Trust). (C) [27] Backyard Ecosystem (Photographer John Castro. Beekeeper Kevin Murphy and bees in urban Denver CO. Photo used courtesy of BackyardEcosystem.com). (D) Landwirtschaftlicher Informationsdienst [28] (© Landwirtschaftlicher Informationsdienst, LID). (E) Who's Robb ? [29] (© Lisa & Robb). (F) Chitra Katha Pvt. [30] (© Sidharto Rao). The ratio a/b (length of the major axis divided by the length of the minor axis) is indicated for each ellipse by a number.

This revealed that two kinds of honeycomb ellipses could be found, according to the values of the ratio provided by dividing the length of the long axis by the length of the short axis of the ellipse. They were found to range either consistently between 1.84 and 2.18 (except for one measure, 2.84 ; n = 28 allowing the generation of a first set of values with a mean 2.0046), or between 1.51 and 1.63 (except for one measure, 1.40 ; n = 16 for the generation of a second set of values with a mean 1.5706). These two sets with all their respective values were further compared together with statistical analyses that are shown in the Table 1. Using Student t-test, the t value was t = 1.42 x 10EE−13. There is 99% probability that the full first set of values falls between the range of 1.9064 and 2.1029 (P <0.01). There is 99% probability that the second full set of values is comprised between 1.5270 and 1.6142 (P <0.01). Finally, the removal of the extreme value 2.84 in the first set of values, and the removal of the minimal value 1.4 in the second set of values, was acceptable for each set of data with a 5% unidirectional risk, as revealed by using the statistical Dixon test. The Shapiro-Wilk normality test performed on both new sets of values revealed that these values were normally distributed (p-value of 0.372 in the first set of values, n = 27; p-value of 0.172 in the second set of values, n = 15). When the full set of data is considered, the mean is equal to 1.847, and the confidence interval (alpha=0.05 ; S.D. of 0.02 ; n=44) is equal to 0.392. Therefore, nothing relevant can be said about the values considered as a whole, thus justufying the generation of two statistically relevant sets of values.

Ratio[†]	1.84 to 2.18	1.51 to 1.63
	[§]2.84	1.63
	2.18	1.63
	2.13	1.63
	2.12	1.63
	2.10	1.60
	2.06	1.59
	2.04	1.59
	2.04	1.58
	2.03	1.58
	2.01	1.58
	2.00	1.57
	2.00	1.55
	1.99	1.53
	1.97	1.53
	1.97	1.51
	1.96	[§]1.40
	1.95	
	1.94	
	1.93	
	1.92	
	1.91	
	1.91	
	1.88	
	1.86	
	1.85	
	1.85	
	1.85	
	1.84	
Mean ± SD[‡]	2.0046 ± 0.1876	1.5706 ± 0.0592
Confidence interval	0.1965	0.0872
Variance	0.034	0.003

Discussion

To my knowledge, there is no systematic mathematical analysis of the ellipses and the ellipsoids formed by honeycombs in their early stages of construction that has been published in the scientific literature, yet. Although the scrutiny of honeycomb pictures in books or electronic sources is providing valuable data, there is unfortunately a lack of reports or scientific articles dealing with the statistical analyses of such ellipsoidal or elliptical honeycombs. Therefore, the comparison of the data presented in this article with the cognate published scientific or beekeeping literature is quite finical.

To date, an abundant and growing scientific literature is focused on the construction of the honeycomb and its hexagonal cells, the nature and production of beeswax, the manipulation of wax by honeybees, the nests and nesting, the self–organisation of nest contents, the wax gland complex and even the repair of experimentally dislocated cells or combs [13,14,31-34], reviewed in [2,3,35].

The ellipsoidal or elliptical forms of honeycombs have been recognised and attested by ancient and more recent authors [2,13,23,36]. My observations suggest that apart from their evident elliptical structure, early stage honeycombs present peculiar intrinsic mathematical properties. When the length of the long axis of the honeycomb ellipse is divided by the length of its minor axis, the resulting ratio is either very close to the number 2.00 or to the number 1.62, with high probabilities. The two rectangles circumscribing these ellipses are revealing the presence of the golden ratio, Φ, with high statistical probabilities.

In the future, similar observations could be obtained and confirmed by other researchers dealing with hives housing honeybees. Will all the early stage honeycombs fit to these observations? What factors are responsible for putative discrepancies? Do honeybee colonies build preferentially honeycombs with one of these two ratios, or both simultaneously, without any preferences? What are the benefits in terms of statics and building behaviour ? Obviously to answer such questions satisfactorily is beyond the scope of this article based on the limited data sets available. A more thorough investigation is therefore needed, in which both the measures of the ellipsoid [37] and the ellipses found in honeycombs might be analysed with procedures involving either real three–dimensional printing or electronic two–dimensional pictures. A real–time investigation in the hive would further allow a precise analysis of the development of the honeycomb.

In order to provide scientific explanations for the understanding of the growing process and the design of the early elliptically–growing honeycomb, one should refer to two specific models. First, the diffusion–limited aggregation (DLA) is a model in which the irreversible morphogenesis of the object arises with scale invariance [10,38]. It includes so–called reaction–diffusion processes which are an essential basis for processes involved in morphogenesis in biology [39].

The second model, the constructal theory, covers natural phenomena of organization and the occurrence of design and patterns in nature based on the laws of physics [11,40,41]. The constructal theory, as a self-standing principle, is distinct from the Second Law of Thermodynamics. Epistemologically, its inherent method proceeds from the simple to the complex; philosophically, the constructal theory "assigns the major role to determinism and contributes significantly to the debate on the origin of living systems" [42,43].

Evidence was presented that "honeybees neither have to measure nor construct the highly regular structures of a honeycomb, and that the observed pattern of combs can be parsimoniously explained by wax flowing in liquid equilibrium. The comb structure is a result of a thermoplastic wax reaching a liquid equilibrium", and that "these interpretations would eliminate the need for bees to perform any mathematical calculations or complex measurements of length and angles" [6]. On the other hand, therefore, we should ask ourselves, whether honeybees do indeed perform calculations involving the golden ratio in order to generate the elliptical honeycombs ! In other words : if the honeybees do not perform these calculations, where and by what or whom are such calculations performed ? This very provocative question encompasses the mathematical, the philosophical, the teleological and/or the spiritual relevance or the ubiquitous presence of the golden ratio in nature [44].

Conclusion

According to the philosopher Immanuel Kant, artworks are made by rational agents : "*For though we like to call the product that bees make (the regularly constructed honeycombs) a work of art, we do so only by virtue of an analogy with art; for as soon as we recall that their labor is not based on any rational deliberation on their part, we say at once that the product is a product of their nature (namely, of instinct)*" [45]. In the years 1850s, Charles Darwin explained the evolution of the honeybee's comb-building abilities, which was essential for the generation of his theory of natural selection . In order to explain the hexagonal geometry of the bee cells, he personnally performed experiments and wrote many letters, the latter being extensively reviewed elsewhere [46]. However, none of these two philosopher or naturalist studied the global elliptical nature of the honeycomb considered as a whole.

The construction of the early honeycomb by honeybees is not a random process, but is following mathematical rules involving the golden ratio. When this honeycomb is presenting an ellipse circumscribed in the long square shown in Fig 5, it reveals a full set of proportions that are related to the golden ratio, as presented in full details elsewhere [47]. When the honeycomb ellipse is circumscribed in the golden rectangle, other interesting properties, also involving the golden ratio, do emerge [48-50].

Fig 5. The double square ellipse and the golden ellipse in honeycombs. Left : a rectangle ABCD having its length AD twice longer as its width AB is called the double square. O is the centre of the rectangle. Numbers refer to the different lengths in arbitrary units. The length of the diagonal AC has the value $\sqrt{5}$. A circle with its centre in O and with a diameter equal to AB is intersecting the diagonal AC in T and U. The distance between the points A and T is equal to the golden ratio Φ = 1.6180339887, approximately (shown in red colour). The distance between the points A and U is equal to $\Phi-1$ = 0.6180339887, approximately. F and F' are the two foci of the ellipse circumscribed in the ABCD rectangle. This ellipse has a ratio (division of the length of the long axis by the length of the short axis of the ellipse) of 2. Its surface is twice the surface of the inner circle. This kind of ellipse is the one that is found in early honeycombs. Right : the golden ellipse comprised within the golden rectangle [48]. An ellipse that would be circumscribed in the golden rectangle having a distance between A and D of 1.6180339887 approximately and a width of 1 is called the golden ellipse. This kind of ellipse is the second one that is found in early honeycombs.

The study of ellipses, especially the golden ellipse and the long square ellipse, is the *parent pauvre* in the observation of the natural world, since mostly spirals, exponentials, polyhedra, pentads, and helicoïdal or symmetrical patterns have been described [51-56]. Mathematics plays a central role in our current scientific picture of the world.

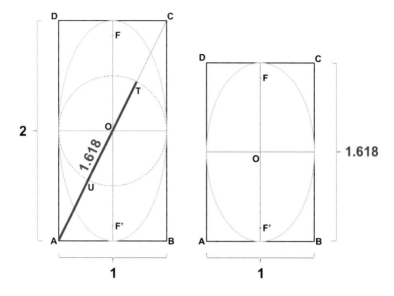

The mathematical explanations in the natural sciences, the explanatory role of mathematics in science and the philosophical relevance of mathematical explanations in science have been reviewed elsewhere :

How the connection between mathematics and the world is to be accounted for remains one of the most challenging problems in philosophy of science, philosophy of mathematics, and general philosophy. A very important aspect of this problem is that of accounting for the explanatory role mathematics seems to play in the account of physical phenomena. [57] The golden ratio is an intrinsic and ubiquitous aspect in such physical phenomena [44].

We should ask ourselves, whether the early phases in the construction of the honeycomb is indeed reflecting the fractal geometry of nature [58]. Fractal structures are abundant in living organisms, ranging from the genetic level, to tissues, organs, organisms and population levels [59]. Moreover, both morphological fractals and temporal fractal structures have been shown to be present within organisms [58]. The growth of the fractal structures has been extensively reviewed not solely for the mineral but also for the animal kingdom [58,60]. Fractal objects do present extremely rich variety of possible realizations of various geometrical objects, to which ellipses might be comprized.

Therefore, the study of the early elliptical honeycomb might thus provide one more note into the hearing of the symphony of life. The fact that the ratios of the honeycomb ellipses do not fit exactly with the values 2.000 or 1.618 can be explained by the observation that in the biological world, living systems are always only approaching their exact mathematical model [53].

The removal of one extreme value in both sets of data is supporting the hypothesis that these two extreme values might belong to a third set of values, namely ellipses circumscribed by rectangles having a ratio which is a multiple of the square root of 2, approximately. Therefore, it is hypothesized that the building of honeycombs by honeybees is following mathematical rules involving irrational numbers, namely the *golden ratio* and possibly the square root of 2.

Further observations around the world will substantiate and confirm these early findings.

Might elliptical honeycombs with geometries close to the *golden ratio* reflect a relatively healthy hive ? Today, a widespread colony collapse disorder is affecting hives worldwide. As written elsewhere by a poet, novelist and nature writer, we have to "create less mechanistic stories about *A. mellifera*" ; in this "alert to the plight of pollinators, writers and artists have begun retelling the bee's story"[61].

The beehive itself is also, figuratively, a microcosm of the biosphere, a concise and comforting poetic image for the architectonics of ecology. Built out of the living substance of bee bodies, the combs of the hive evoke, in their intricate cell-structure, the architecture of niches that characterizes the biosphere [62]. This architecture is the revelation and the representation of an ancient knowledge, called gnomonicity, related to the word gnomon [63]. The word gnomon was originally given by the Hero of Alexandria, such as "the form that, when added to some form, results to a new form similar to the original". The addition of successive geometric gnomons to a figure does not alter its proportions, as it is the case for the early honeycomb (Fig 6). The mathematical and scientific explanation of the gnomonicity is thus avoiding the pitfalls of the widespread and unfortunate golden ratio mysticism.

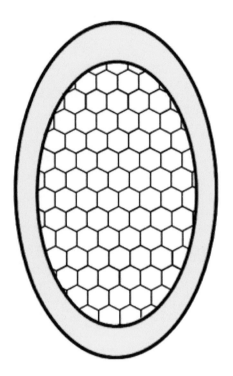

Fig 6. The gnomonic growth of the early honeycomb. The elliptical gnomon is shown in yellow. It is resembling an "elliptical donut".

Acknowledgements

This research was performed under the full responsibility of the main author (D.F.) I thank Prof. Harald Berresheim for helpful comments and for the critical reading of the manuscript, Lawrence Michel for his advice in the statistical analyses, and Léonard Ribordy for his comment regarding the golden ellipse.This article is dedicated to the memory of the late mathematician Midhat J. Gazalé.

References

1. Hepburn HR, Pirk CWW, Duangphakdee O. Honeybee Nests: Composition, Structure, Function. Berlin Heidelberg: Springer-Verlag; 2014.

2. Hepburn HR. Honeybees and wax : an experimental natural history. Berlin: Springer-Verlag; 1986.

3. Hepburn HR, Pirk CWW, Duangphakdee O. Construction of Combs. Honeybee Nests: Berlin Heidelberg: Springer; 2014. pp. 207-221.

4. Yang MX, Tan K, Radloff SE, Phiancharoen M, Hepburn HR. Comb construction in mixed-species colonies of honeybees, Apis cerana and Apis mellifera. J Exp Biol. 2010; 213: 1659-1664.

5. Beekman M, Sword G, Simpson S. Biological foundations of swarm intelligence. In: Swarm intelligence: Blum, C. Merkle, D, editors. Heidelberg; 2008. pp. 3-41.

6. Pirk CWW, Hepburn HR, Radloff SE, Tautz J. Honeybee combs: construction through a liquid equilibrium process? Naturwissenschaften 2004 ; 91: 350-353.

7. Karihaloo BL, Zhang K, Wang J. Honeybee combs: how the circular cells transform into rounded hexagons. J R Soc Interface 2013; 10: 20130299.

8. Belić MR, Škarka V, Deneubourg JL, Lax M (1986) Mathematical model of honeycomb construction. J Math Biol. 24: 437-449.

9. Baldwin SA. Some historical aspects of honeycomb mathematics. Bee Craft Ltd - The Informed Voice of British Beekeeping. 2012; 27-30.

10. Witten TA, Sander LM. Diffusion-limited aggregation. Phys Rev. B 1983; 27: 5686-5697.

11. Bejan A, Lorente S. Constructal theory of generation of configuration in nature and engineering. J Applied Physics. 2006; 100: 041301.

12. Murthy A (2013) Mathematics of nature and nature of mathematics. Chapter I. Available : http://www scribdcom/doc/21990600/Maths-of-Nature-and-Nature-of Maths. Accessed 1 January 2016.

13. Darchen R. Les techniques de construction chez "Apis mellifica": Masson. 1959.

14. Darchen R. Construction et reconstruction de la cellule des rayons d'Apis mellifica. Insectes Sociaux 1958; 5: 357-371.

15. Wikipedia : Golden ratio. Available : https://en.wikipedia.org/wiki/Golden_ratio. Accessed 1 January 2016.

16. Hemenway P. Divine Proportion: Phi in Art, Nature, and Science. New York: Sterling Publishing Company; 2005.

17. Ghyka MC. Esthétique des proportions dans la nature et dans les arts: Géométrie des formes naturelles inorganiques et vivantes. Paris: Gallimard; 1927.

18. Abeille&Nature : Les cadres à jambage pour la ruche. Le Cergne, France. Available : http://www.abeille-et-nature.com/index.php?cat=apiculture&page=cadres_a_jambage. Accessed 1 January 2016.

19. Agroscope Liebefeld-Posieux, Centre de Recherches Apicoles, Berne, Switzerland. Available : http://www.agroscope.admin.ch/imkerei/index.html?lang=en. Accessed 1 January 2016.

20. Charrière J-D, Imdorf A, Kuhn R. Tolérance pour les abeilles de différents traitements hivernaux contre Varroa. Journal Suisse d'Apiculture. 2004; 125(5). Available : https://www.google.ch/url?sa=t&rct=j&q=&esrc=s&source=web&cd=1&cad=rja&uact=8&ved=0ahUKEwjkq76duIvKAhUFaxQKHQ_zAFUQFggjMAA&url=http%3A%2F%2Fwww.agroscope.admin.ch%2Fimkerei%2F00316%2F00329%2F02081%2Findex.html%3Flang%3Dfr%26download%3DNHzLpZeg7t%2Clnp6IoNTU042l2Z6ln1ae2IZn4Z2qZpnO2Yuq2Z6gpJCDeHt%2Cfmym162epYbg2c_JjKbNoKSn6A--&usg=AFQjCNESicENa2iLYRUNlOIsqQ96F6FqTg&bvm=bv.110151844,d.bGQ. Accessed 1 January 2016.

21. Wikipedia : Ellipse. Available : https://en.wikipedia.org/wiki/Ellipse. Accessed 1 January 2016.

22. Hepburn H, Whiffler L. Construction defects define pattern and method in comb building by honeybees. Apidologie. 1991; 22: 381-388.

23. Layens dG, Bonnier G. Cours complet d'apiculture (Culture des Abeilles); Dupont P, editor (Paris). 1897.

24. Hepburn HR, Pirk CWW, Duangphakdee O. Intraspecific and Interspecific Comb-Building. Honeybee Nests.: Berlin Heidelberg: Springer; 2014. pp. 57-78.

25. Ottawa Honey House. The experiment begins. Available : http://ottawahoneyhouse.com/2014/06/30/the-experiment-begins/. Accessed 1 January 2016.

26. Natural Beekeeping Trust : – Sun Hive Bees. Available : https://naturalbeekeepingtrust.files.wordpress.com/2015/02/natural-bee.jpg. Accessed 1 January 2016.

27. Backyard Ecosystem: So exactly what is natural comb anyway ? Available : http://www.backyardecosystem.com/wp-content/uploads/2011/03/newfoundation-e1299117039501.jpg. Accessed 1 January 2016.

28. Landwirtschaftlicher Informationsdienst : Bienen Wichtige Helferinnen der Schweizer Bauern. Bern, Switzerland. Available : http://www.lid.ch/fileadmin/user_upload/lid/Produkte/Broschueren/20392d_Broschuere_Bienen.pdf. Accessed 1 January 2016.

29. Who's Robb ?: A hive of activity. Available : http://howsrobb.blogspot.ch/2010/05/hive-of-activity.html. Accessed 1 January 2016.

30. Chitra Katha Pvt. Ltd Brainwave: Science is just a game! Available : http://www.bwmag.in/wp-content/uploads/2011/12/honey-comb.jpg. Accessed 1 January 2016.

31. Seeley TD, Morse RA. The nest of the honey bee (Apis mellifera L.). Insectes Sociaux. 1976; 23: 495-512.

32. Darchen R. Le travail de la cire et la construction dans la ruche. In: Chauvin R, editor. Traité de biologie de l'abeille. Paris: Masson; 1968. pp. 252-331.

33. Frisch Kv. Animal architecture. 1st ed. New York, Harcourt Brace Jovanovich; 1974.

34. Werner-Meyer W. Wachs und Wachsbau - Kittharz. In: Büdel H, editor. Biene und Bienenzucht. München: Ehrenwirth Verlag. 1960; pp. 202-232.

35. Tóth LF. What the bees know and what they do not know. Bull Amer Math Soc. 1964; 70: 468-481.

36. Huber F (1814) Nouvelles observations sur les abeilles, tome second. Available : http://www.e-rara.ch/zut/content/structure/7077997. Accessed 1 January 2016.

37. Kapur JN. The golden ellipsoids and golden hyperellipsoids. Internat J Math Ed Sci Tech. 1987; 18: 699-704.

38. Sander LM. Diffusion-limited aggregation: A kinetic critical phenomenon? . Contemp Phys. 2000; 41: 203-218.

39. Stevens PS. Patterns in Nature. 1st ed. Boston: Little, Brown and Company; 1974.

40. Bejan A, Lorente S. Design with Constructal Theory. Int J Eng Edu. 2006; 22: 140-147.

41. Bejan A, Lorente S. The constructal law of design and evolution in nature. Phil Trans R Soc B. 2010; 365: 1335-1347.

42. Reis A. Constructal Theory: From Engineering to Physics, and How Flow Systems Develop Shape and Structure. Appl Mech Rev. 2006; 59: 269-282.

43. Schrödinger E. What is Life ? The Physical Aspect of the Living Cell. Cambridge University Press; 1944.

44. Boeyens JCA, Thackeray JF. Number theory and the unity of science. S Afr J Sci. 2014; Art. #a0084. Available : http://www.sajs.co.za/sites/default/files/publications/pdf/Boeyens_SciCo.pdf. Accessed 1 january 2016.

45. Kant I. Critik der Urtheilskraft (Critique of the Power of Judgment). Berlin und Lidau; 1790. Available : http://oll.libertyfund.org/titles/1217. Accessed 1 january 2016.

46. Burkhardt F (2015) The evolution of honeycomb. Darwin Correspondence Project. University of Cambridge, UK. Available : https://www.darwinproject.ac.uk/the-evolution-of-honey-comb. Accessed 1 January 2016.

47. Koelliker T. Symbolisme et Nombre d'Or - Le rectangle de la Genèse et la Pyramide de Khéops, avec fascicule-annexe de 48 figures. Paris: Les Editions des Champs-Elysées ; 1957.

48. Kapur JN. The golden ellipse. Internat J Math Ed Sci Tech. 1987; 18: 205-214.

49. Kapur JN. The golden ellipse revisited. Internat J Math Ed Sci Tech. 1988; 19: 787-793.

50. Huntley HE. The golden ellipse. The Fibonacci Quarterly. 1974; 12: 38-40.

51. Thompson DAW. On Growth and Form. New York: The MacMillan Company; 1942.

52. Adam JA. Mathematics in Nature: Modeling Patterns in the Natural World. Princeton: Princeton University Press; 2003.

53. Cook TA. The Curves of Life: Being an Account of Spiral Formations and Their Application to Growth in Nature, to Science, and to Art, with Special Reference to the Manuscripts of Leonardo Da Vinci. Mineola, New York: Dover Publications; 1979

54. Colman S. Proportional Form: Further studies in the science of beauty, being supplemental to those set forth in "Nature's harmonic Unity." By Samuel Colman and C. Arthur Coan. The drawings and correlating descriptions are by Colman, the text and mathematics are by Coan: G. P. New York: Putnam's Sons; 1920.

55. Weyl H. Symmetry. Princeton: Princeton University Press; 1952.

56. French KL (2014) The Hidden Geometry of Life: The Science and Spirituality of Nature. London: Watkins Media Limited; 2014.

57. Mancosu P (2008) Explanation in Mathematics. The Stanford Encyclopedia of Philosophy. Zalta EN, editor. Available : http://plato.stanford.edu/archives/sum2015/entries/mathematics-explanation/. Accessed 1 January 2016.

58. Queiros-Condé D, Chaline J, Méhauté AL, Dubois J. Le monde des fractales: La nature trans-échelles. Paris: Ellipses Marketing; 2015.

59. Posamentier AS, Lehmann I. The Glorious Golden Ratio. Amherst, New York: Prometheus Books; 2012.

60. Fleury V. Arbres de pierre: la croissance fractale de la matière: Paris: Flammarion; 1998.

61. Burnside J. Apiculture: Telling the bees. Nature. 2015; 521: 29-30.

62. Mathews F. Planet Beehive. Australian Humanities Review. 2011; 50: 159-178. Available : http://www.australianhumanitiesreview.org/archive/Issue-May-2011/mathews.html. Accessed 1 January 2016.

63. Gazalé MJ. Gnomon: From Pharaohs to Fractals: Princeton University Press ; 1999.

Addendum

The scientific article entitled "Golden ratio (Sectio aurea; divine proportion) in the early elliptical honeycomb of Apis mellifera" was submitted thrice for publication in peer-reviewed scientific journals. It was rejected for the following reasons :

Journal of ********** ********
"I regret not paying enough attention to both the methods and the discussion before sending this out for review. I think both have serious problems, some of which are addressed on an attached file. There are numerous problems with the English in spelling, grammar and word choice but these are not disqualifiers.

A major problem is a failure to explain how the two means of the ratio of the dimensions of the combs (given the reasonable assumption that they are ellipses) "reveals" the presence of the golden ratio or even why the distribution of the means should be bimodal. Even given the assumption that the golden ratio pertains here, any connection with biology is quite tenuous. The author refers to a "hidden" golden ratio but this seems like a non-sequtur. The very extensive additions to the revised manuscript did little to clarify things and most cases simply muddy the waters. Much seems to wander off into obscure philosophical byways. As such, I don't think the paper is appropriate for ***."

Journal of *******
"The article discusses the issue of two-dimessional form of a honey bee comb at the early stages of its formation. In particularly it aims to state that its construction follows mathematical rules. However, no solid data are presented to support this conclusion. More data are needed to be presented to support the hipothesis. These also need to be evaluated by more than discriptive statistics. The data are not clearly explanation connecting between 1.6 and 2.0 is given. Significance of this reserch is not clearly stated."

**** ***

"Thank you for submitting your manuscript to **** ***. After careful consideration, we have decided that your manuscript does not meet our criteria for publication and must therefore be rejected. Specifically:
The paper aims to present evidence for the presence of

the golden ratio in the elliptical honeycomb of Apis mellifera, but in fact the majority of the values presented fall well outside of this range (i.e. the dataset 1.84 to 2.18). The relevance of this second range is not discussed.

Even if clear evidence for the golden ratio were given, the presented finding would only be of scientific interest if a mechanistic explanation were given: i.e., an explanatory, mechanistic physical model. The author mentions fractal models, e.g., diffusion-limited aggregation, but the presented elliptical shapes of honeycombs are clearly not fractal. A clear physical or biological model explaining the finding is not presented; instead, based on poor numerical evidence, the author seems to point at super-natural explanations for the presumed presence of the golden ratio in early honeycombs: "if the honeybees do not perform these calculations, where and by what or whom are such calculations performed ? This very provocative question encompasses the mathematical, the philosophical, the teleological and/or the spiritual relevance or the ubiquitous presence of the golden ratio in nature [44]".

I am sorry that we cannot be more positive on this occasion, but hope that you appreciate the reasons for this decision."

CPSIA information can be obtained at www.ICGtesting.com
Printed in the USA
LVIW01n1124261216
518710LV00006B/11